THE POETRY OF PRASEODYMIUM

The Poetry of Praseodymium

Walter the Educator

Silent King Books a WhichHead imprint

Copyright © 2023 by Walter the Educator

All rights reserved. No part of this book may be reproduced in any manner whatsoever without written permission except in the case of brief quotations embodied in critical articles and reviews.

First Printing, 2023

Disclaimer
This book is a literary work; poems are not about specific persons, locations, situations, and/or circumstances unless mentioned in a historical context. This book is for entertainment and informational purposes only. The author and publisher offer this information without warranties expressed or implied. No matter the grounds, neither the author nor the publisher will be accountable for any losses, injuries, or other damages caused by the reader's use of this book. The use of this book acknowledges an understanding and acceptance of this disclaimer.

"Earning a degree in chemistry changed my life!"
- Walter the Educator

dedicated to all the chemistry lovers, like myself, across the world

CONTENTS

Dedication v

Why I Created This Book? 1

One - Element So Grand 2

Two - Oh, Praseodymium 4

Three - Symbol Of Wonder 6

Four - Masterpiece Of Art 8

Five - Praseodymium, A Treasure 10

Six - Chemistry Sea 12

Seven - Fame 14

Eight - Boldly Declare 16

Nine - Etched In Our Heart 18

Ten - Unique Charm 20

Eleven - Forever It Presides 22

Twelve - Bright And Stable 24

Thirteen - Eternal Fable	. . . . .	26
Fourteen - We Celebrate	. . . . .	28
Fifteen - Reign Supreme	. . . . .	30
Sixteen - Hearts Ponder	. . . . .	32
Seventeen - Scientific Light	. . . . .	34
Eighteen - Scientific History	. . . . .	36
Nineteen - Symbol Of Curiosity	. . . . .	38
Twenty - Curiosity	. . . . .	40
Twenty-One - World Becomes Aware	. . . .	42
Twenty-Two - Enigmatic Key	. . . . .	44
Twenty-Three - Within Thee	. . . . .	46
Twenty-Four - Secrets Of Chemistry	. . . . .	48
Twenty-Five - Element With Soul	. . . . .	50
Twenty-Six - Element Of Allure	. . . . .	52
Twenty-Seven - Puzzle To Explore	. . . . .	54
Twenty-Eight - Divine	. . . . .	56
Twenty-Nine - Praseodymium, An Element Revered	. . . . .	58
Thirty - Scientific Desire	. . . . .	60
Thirty-One - Intriguing	. . . . .	62

Thirty-Two - Every Discovery	. . . .	64
Thirty-Three - Zest	. . . .	66
Thirty-Four - Forever Inspiring	. . . .	68
Thirty-Five - Scientific Ways	. . . .	70
About The Author	. . . .	72

WHY I CREATED THIS BOOK?

Creating a poetry book about the chemical element of Praseodymium was an intriguing and unique task. Praseodymium, with its distinct properties and characteristics, can serve as a source of inspiration. By exploring its atomic structure, history, and applications in various fields, I can delve into themes of transformation, rarity, and the interconnectedness of science and art. This poetry book can offer readers a fresh perspective on the world of chemistry, bridging the gap between science and creativity, and stimulating curiosity about the elements that surround us.

ONE

ELEMENT SO GRAND

In the realm of elements, a gem does shine,
Praseodymium, a name quite divine.
With atomic number fifty-nine it stands,
A symbol of rarity in far-off lands.

    A lustrous metal, its hue is distinct,
A gentle greenish-yellow, I think.
Its beauty, like a meadow bathed in light,
Captivating all with its radiant might.

    Praseodymium, a treasure to behold,
In the tales of chemistry, often untold.
Its presence in alloys, a secret blend,
Enhancing strength, a helping hand to lend.

    This element, with its magnetic might,
Draws us closer, like the stars at night.

It dances within our technology,
Powering progress with its chemistry.
 Its name derived from the Greek, "prasios",
Meaning "leek green," a color that bestows
A sense of life, of growth, and renewal,
Praseodymium, an element so crucial.
 So let us celebrate this rare delight,
Praseodymium, a beacon in the night.
In the vast tapestry of elements grand,
It shines with brilliance, an element so grand.

TWO

OH, PRASEODYMIUM

In the kingdom of elements, behold Praseodymium,
A shimmering gem, a marvel beyond the spectrum.
With atomic number fifty-nine, it reigns supreme,
A luminescent tale, a scientist's dream.
    Its essence, a symphony of colors untamed,
From pale green to golden, a palette unrestrained.
Praseodymium, a chameleon of the periodic chart,
Radiating beauty, igniting the poet's heart.
    In rare earth minerals, it finds its sacred space,
A conductor of energy, an alchemist's embrace.
A dashing partner in alloys, it lends its might,
Infusing strength, forging bonds with sheer delight.
    Praseodymium, a guardian of magnetic forces,
Guiding compass needles, unveiling hidden courses.
Its magnetic dance, an enigma, a cosmic affair,
Harnessing the power of attraction, beyond compare.

Within the realm of technology, it plays its part,
A silent ally, a catalyst for innovation to start.
From lasers to glass, its touch is pure and true,
Lighting the path to progress, unveiling the new.

Oh, Praseodymium, a treasure of the Earth's embrace,
A testament to nature's artistry, woven with grace.
In the grand tapestry of elements, you shine bright,
Praseodymium, an element of wonder, pure and light.

THREE

SYMBOL OF WONDER

In the realm of chemistry, a jewel does appear,
Praseodymium, a name that whispers in the ear.
With atomic number fifty-nine, it claims its throne,
A mystical element, a mystery to be known.

A luminescent green, its color does ignite,
A flickering flame, a beacon shining so bright.
Praseodymium, a secret to be unfurled,
A tale of rare beauty, in the science world.

Within rare earth minerals, it finds its home,
A hidden treasure, waiting to be known.
Its presence in magnets, a force to behold,
Attracting attention, with stories untold.

Praseodymium, a conductor of light,
Radiating brilliance, dispelling the night.
From lasers to glass, its touch is profound,
Unveiling the wonders, in discoveries found.

In the realm of alloys, it lends its strength,
A silent partner, in the forging of great lengths.
Enhancing properties, with its subtle touch,
Praseodymium, a catalyst, we cherish so much.

Oh, Praseodymium, a gem of the periodic chart,
A symbol of wonder, a treasure to impart.
In the realm of elements, you hold your sway,
Praseodymium, we celebrate you today.

FOUR

MASTERPIECE OF ART

In the realm of elements, a tale unfolds,
Of Praseodymium, a story yet untold.
With its atomic number, fifty-nine it claims,
A luminescent gem, dancing in vibrant flames.

An emerald hue, it dazzles like no other,
Praseodymium, a rare and precious smother.
Within its bonds lie secrets yet unveiled,
A symphony of chemistry, where wonders are hailed.

In the depths of Earth, it finds its dwelling,
Nestled in minerals, silently compelling.
A guardian of magnets, with magnetic might,
Praseodymium, drawing forces, day and night.

In alloys it blends, a catalyst profound,
Transforming the mundane, with strength renowned.

A touch of magic, enhancing its core,
Praseodymium, an element to adore.
With lasers, it dances, a mesmerizing show,
Unveiling the mysteries, where knowledge will grow.
In glass it weaves, a tapestry of light,
Illuminating truth, banishing the night.
Oh, Praseodymium, a gem of allure,
A luminary element, forever pure.
In the realm of science, you claim your space,
Praseodymium, an emblem of grace.
So let us celebrate this element rare,
Praseodymium, with its wonders to share.
In the vast symphony of the periodic chart,
Praseodymium, a masterpiece of art.

FIVE

PRASEODYMIUM, A TREASURE

In the realm of elements, a gem so rare,
Praseodymium, with its vibrant flair.
From the depths of Earth, it emerges bright,
Igniting curiosity, a celestial light.
    A hue like no other, a beguiling shade,
A dance of greens, a chromatic cascade.
Praseodymium, a painter's delight,
Brushing the canvas with colors so bright.
    Within alloys, it weaves a tale,
Bestowing strength, an armor to prevail.
A secret ingredient, a touch of grace,
Praseodymium, enhancing with its embrace.
    In magnets it thrives, a magnetic force,
Guiding the compass, staying on course.

With its magnetic marvel, it captivates,
Praseodymium, the element that fascinates.
In lasers, it shines a focused beam,
Unveiling the unseen, like a lucid dream.
A symphony of light, an intricate play,
Praseodymium, illuminating the way.
Oh, Praseodymium, a jewel rare and fine,
A testament to nature's design.
In the grand tapestry of elements, you shine,
Praseodymium, a treasure, forever divine.

SIX

CHEMISTRY SEA

In the realm of elements, Praseodymium stands,
A whispering secret, held in nature's hands.
With a hue of green, like a leaf in the breeze,
It dances with grace, on the periodic seas.

Praseodymium, a creator of light,
In lasers it sparkles, a dazzling sight.
Unveiling the secrets, hidden in the dark,
It paints a path, with its luminous arc.

In alloys, it melds, a magical touch,
Binding the elements, with strength as such.
A silent conductor, orchestrating the blend,
Praseodymium, a harmonious friend.

Magnetic allure, it possesses with glee,
Guiding the compass, a navigator's key.

A force in the field, attracting the eye,
Praseodymium, a magnetism we can't deny.
   Oh, Praseodymium, an element of wonder,
In the symphony of atoms, you strike like thunder.
A guardian of light, a catalyst to be,
Praseodymium, a treasure in the chemistry sea.

SEVEN

FAME

In the realm of chemistry, a gem so rare,
Praseodymium, with mysteries to share.
With atomic beauty, it graces the stage,
A beacon of wonder, an element's sage.

In alloys, it weaves a tapestry strong,
A secret ingredient, where metals belong.
Enhancing their power, creating alloys bold,
Praseodymium, a catalyst of stories untold.

Its color, a spectacle, like nature's hue,
Leek-green radiance, a sight to imbue.
In the spectrum of light, it dances with grace,
Praseodymium, painting the world's embrace.

With magnetic charm, it captivates the field,
Guiding compass needles, its influence revealed.

A force of attraction, it holds us tight,
Praseodymium, a magnetic delight.
   Oh, Praseodymium, element of allure,
In the grand tapestry of elements, pure.
A symbol of power, a catalyst's name,
Praseodymium, forever in chemistry's fame.

EIGHT

BOLDLY DECLARE

A gem in disguise, Praseodymium shines,
With secrets and wonders, it intertwines.
In the palette of elements, it claims its space,
Praseodymium, a symbol of grace.

    A conductor of light, it dances and gleams,
Illuminating paths, like radiant beams.
From lasers it springs, a vibrant display,
Praseodymium, lighting the way.

    In glass it resides, a touch of allure,
Adding brilliance, like a spell to ensure.
A prism of colors, a kaleidoscope,
Praseodymium, enchanting our scope.

    Oh, Praseodymium, element of mystery,
Unveiling truths, with a touch of history.

In the grand tapestry of chemistry's art,
Praseodymium, a masterpiece, a part.
    With magnetism it plays, a captivating force,
Guiding our compass, staying on course.
Oh, Praseodymium, element so rare,
In the realm of elements, you boldly declare.

NINE

ETCHED IN OUR HEART

Praseodymium, a touch of celestial grace,
An element that paints the cosmos' face.
In the depths of the universe, it shines,
A stellar presence, among celestial signs.

With a hue of green, like the enchanted earth,
Praseodymium, a symbol of renewal and rebirth.
In crystals and glass, it weaves its spell,
A prism of colors, where wonders dwell.

Magnetic allure, it holds in its core,
Guiding the compass, a navigator's lore.
A force of attraction, both strong and true,
Praseodymium, a magnetic marvel for us to view.

Oh, Praseodymium, element of cosmic dance,
In the vast expanse, your brilliance enchants.

A catalyst of light, in lasers it thrives,
Unveiling secrets, as the universe strives.
    In the pantheon of elements, you claim your place,
Praseodymium, a celestial embrace.
A testament to nature's alchemical art,
Praseodymium, forever etched in our heart.

TEN

UNIQUE CHARM

Praseodymium, an element so fine,
Within the periodic table, you brightly shine.
With atomic number fifty-nine,
A rare gem in the chemical line.
    In the realm of magnets, you wield your might,
A magnetic force, captivating our sight.
Guiding compass needles, true and steady,
Praseodymium, a magnetic symphony ready.
    In the kingdom of light, you cast your glow,
With lasers, you dance, a mesmerizing show.
Illuminating the path, with radiant hues,
Praseodymium, a luminary muse.
    A catalyst of wonders, you do bestow,
In chemical reactions, your power does grow.

Enhancing alloys, creating strength anew,
Praseodymium, a catalyst so true.
 Oh Praseodymium, element divine,
A treasure in nature's design.
With your unique charm, you captivate us all,
Praseodymium, forever we shall recall.

ELEVEN

FOREVER IT PRESIDES

Praseodymium, an element of grace,
With an emerald glow, it lights up space.
In the realm of chemistry, it holds its own,
A marvel of nature, beautifully shown.

With magnetic strength, it pulls the strings,
Guiding compasses, like celestial wings.
A force of attraction, a magnetic embrace,
Praseodymium, leaving no trace.

In lasers, it dances, a vibrant display,
Unleashing energy in a captivating way.
With its green brilliance, it paints the night,
Praseodymium, a beacon of light.

Oh Praseodymium, element of allure,
In the periodic table, you shine pure.

A catalyst of wonders, bringing change,
Praseodymium, in the chemistry range.
 In alloys, it blends, adding strength and might,
Transforming materials, like a shining knight.
Praseodymium, a symbol of transformation,
A testament to its elemental creation.
 So let us marvel at this rare gem,
Praseodymium, an element to condemn.
For in its essence, beauty resides,
Praseodymium, forever it presides.

TWELVE

BRIGHT AND STABLE

In the realm of elements, a gem so rare,
Praseodymium, with secrets to share.
A symphony of electrons, dancing in delight,
Unveiling the wonders of its atomic flight.

With a vibrant hue, like nature's green,
Praseodymium, a sight to be seen.
In the glass it dwells, a touch of grace,
Adding allure to its crystal embrace.

Magnetic forces, it wields with finesse,
Guiding compass needles, no need to guess.
A conductor of attraction, pulling us near,
Praseodymium, a magnetic pioneer.

Oh, Praseodymium, element of awe,
In the chemistry world, you leave us in awe.

A catalyst of light, in lasers you thrive,
Revealing the secrets that make us alive.

In the grand tapestry of the periodic table,
Praseodymium, you shine bright and stable.
A symbol of wonder, a treasure untold,
Praseodymium, a marvel to behold.

THIRTEEN

ETERNAL FABLE

Praseodymium, element of rare delight,
In the realm of chemistry, a radiant light.
With a vibrant hue, green as the spring,
You captivate our senses, a magical thing.
 Magnetic enchantress, you hold a force,
Guiding compass needles, staying on course.
A magnetic marvel, a conductor of power,
Praseodymium, in your presence we cower.
 In the heart of lasers, you gracefully dance,
Creating beams of brilliance, a cosmic romance.
With your atomic prowess, you paint the sky,
Praseodymium, a celestial lullaby.
 A catalyst of change, you work unseen,
Transforming reactions, like alchemy's dream.

In alloys you mingle, enhancing their might,
Praseodymium, an element shining bright.
  Oh Praseodymium, your name we revere,
A symbol of wonder, forever held dear.
In the vast expanse of the periodic table,
Praseodymium, you shine, an eternal fable.

FOURTEEN

WE CELEBRATE

In the realm of elements, you hold your place,
Praseodymium, with elegance and grace.
A symphony of electrons, dancing in sync,
Unleashing your power, refusing to shrink.

With a hue of green, like a precious gem,
Praseodymium, you captivate and stem
The flow of time, as you shimmer and gleam,
A luminescent beauty, a spectral dream.

In magnets, you hold the force of attraction,
Guiding us with precision, a flawless reaction.
Praseodymium, conductor of the unseen,
Weaving magnetic fields, an enigmatic routine.

A catalyst of change, in chemistry's domain,
You enhance the mixtures, a transformative chain.

Praseodymium, catalyst of alchemical might,
Unveiling the secrets hidden from sight.
    Oh Praseodymium, element of wonder,
In your atomic realm, we often ponder
The mysteries you hold, the magic you create,
Praseodymium, forever we celebrate.

FIFTEEN

REIGN SUPREME

Praseodymium, a gem of rare sight,
In the world of elements, you shine bright.
With hues of green, an enchanting hue,
A symphony of atoms, a remarkable debut.

    Oh Praseodymium, conductor of fire,
In lasers, you dance, igniting desire.
With precision and focus, you guide the way,
A beacon of light, leading us astray.

    In alloys, you blend, a transformative blend,
Enhancing strength, a metamorphosis you lend.
Praseodymium, catalyst of metallurgical might,
Unveiling the possibilities, an alchemical flight.

    Magnetic fields bend and bow in your presence,
Compass needles sway, drawn to your essence.

Praseodymium, wielder of magnetic forces,
A symphony of attraction, nature's courses.
   Oh Praseodymium, element of allure,
Your atomic dance, so pure and secure.
In the realm of chemistry, you reign supreme,
Praseodymium, a testament to the unseen.

SIXTEEN

HEARTS PONDER

In the realm of elements, a gem so rare,
Praseodymium, with an aura beyond compare.
A symphony of hues, a palette divine,
In shades of green, your brilliance shines.

    A conductor of light, in lasers you dwell,
Harnessing photons, a magical spell.
Praseodymium, illuminating the night,
With spectral beams, a celestial delight.

    In the heart of magnets, you hold the key,
Guiding forces, with magnetic decree.
A dance of poles, a magnetic embrace,
Praseodymium, enchanting in every space.

    A catalyst of change, a catalyst of might,
In chemical reactions, you ignite.

Transforming compounds, with a subtle touch,
Praseodymium, catalyst of alchemical clutch.
  Oh Praseodymium, element of wonder,
In the periodic table, you make hearts ponder.
With your atomic allure, a beauty untold,
Praseodymium, a treasure to behold.

SEVENTEEN

SCIENTIFIC LIGHT

Praseodymium, oh element of grace,
With your presence, the world finds its place.
In the realm of science, you hold a key,
Unveiling secrets, for all to see.

A hue of green, like the forest's canopy,
Praseodymium, a shade of mystery.
In lasers, you dance, emitting light,
A symphony of colors, shining bright.

Magnetic marvel, conductor supreme,
Pulling us closer, like a magnetic dream.
In compass needles, you point the way,
Guiding explorers, both night and day.

A catalyst of change, you alter the course,
Transforming reactions with your force.
Praseodymium, catalyst of alchemical might,
Unlocking the potential, shining so bright.

Oh Praseodymium, element divine,
In the periodic table, you truly shine.
With your atomic charm, a captivating sight,
Praseodymium, a beacon of scientific light.

EIGHTEEN

SCIENTIFIC HISTORY

Praseodymium, an element of grace,
In the realm of chemistry, you find your place.
With atomic number fifty-nine,
You shimmer and glow, like a celestial sign.

In the depths of the Earth, where minerals reside,
You hide in abundance, a treasure to confide.
With hues of green, like emerald's gleam,
Praseodymium, a gem in nature's dream.

Magnetic fields bend and sway,
In your presence, they cannot disobey.
A conductor of power, you lead the way,
Praseodymium, guiding us day by day.

A catalyst of change, you spark the flame,
Transforming compounds, never the same.

With your alchemical touch, reactions unfold,
Praseodymium, a secret to be told.

Oh Praseodymium, element of allure,
In the world of science, you endure.
With your atomic charm, so rare and true,
Praseodymium, we marvel at you.

In the vast expanse of the periodic table,
You stand tall, an element that is stable.
Praseodymium, a symbol of discovery,
Forever celebrated in scientific history.

NINETEEN

SYMBOL OF CURIOSITY

Praseodymium, rare and sublime,
An element of wonder, frozen in time.
In the realm of atoms, you hold your ground,
With a vibrant hue, like a gemstone found.

Magnetic marvel, conductor of might,
Guiding forces with magnetic insight.
In magnets you dwell, creating a force,
Praseodymium, magnetic to the core.

A catalyst of change, you spark the reaction,
Transforming compounds with precise interaction.
Praseodymium, catalyst of alchemical art,
Unveiling the secrets, playing your part.

Oh Praseodymium, element of grace,
In the periodic table, you find your place.

With atomic allure, you captivate,
Praseodymium, an element we celebrate.
   In lasers you shine, illuminating the way,
Casting beams of brilliance, a mesmerizing display.
Praseodymium, a luminescent delight,
Guiding us through darkness, with your radiant light.
   Oh Praseodymium, element divine,
In the world of science, you forever shine.
With your unique properties, a marvel to see,
Praseodymium, a symbol of curiosity.

TWENTY

CURIOSITY

Praseodymium, a gem of the atomic realm,
With hues of green, a mystical helm.
In magnetic fields, you hold the sway,
Aligning the forces, in a symphony you play.
Praseodymium, conductor of attraction,
Weaving magnetic tapestries, a cosmic interaction.
A catalyst of change, in the realm of chemistry,
You bring transformations, a catalyst of mystery.
Praseodymium, alchemical artisan,
Unveiling the secrets, a magician in the plan.
Oh Praseodymium, element of wonder,
In the periodic table, you truly ponder.
With your atomic dance, a mesmerizing sight,
Praseodymium, shining with celestial light.
In lasers, you shimmer, a radiant glow,

Guiding the path, where energy does flow.
Praseodymium, illuminating our way,
A beacon of knowledge, forever we'll stay.
    Oh Praseodymium, element so rare,
With your presence, the world becomes aware.
A symbol of curiosity, in science you thrive,
Praseodymium, forever you'll strive.

TWENTY-ONE

WORLD BECOMES AWARE

In the realm of elements, you hold your place,
Praseodymium, an enigma to embrace.
With atomic grace, you shimmer and gleam,
A symphony of electrons, like a vibrant dream.

Oh Praseodymium, element of intrigue,
In the periodic table, you truly intrigue.
With your magnetic charm, a captivating force,
Praseodymium, guiding our scientific course.

In the heart of lasers, you dance and play,
Emitting hues of green, a dazzling display.
Praseodymium, conductor of light,
Leading us through darkness, with colors so bright.

A catalyst of change, you ignite the spark,
Transforming reactions in the deepest dark.

Praseodymium, alchemical alibi,
Unveiling the secrets, as time goes by.
   Oh Praseodymium, element so rare,
With your presence, the world becomes aware.
A symbol of curiosity, in science you thrive,
Praseodymium, forever you'll strive.

TWENTY-TWO

ENIGMATIC KEY

Praseodymium, a jewel of the periodic table,
A treasure of science, a story to unravel.
With atomic grace, you capture our gaze,
Praseodymium, in your vibrant ways.

In the realm of magnets, you hold the sway,
Guiding forces, in your magnetic ballet.
A symphony of poles, a magnetic embrace,
Praseodymium, with elegance and grace.

A catalyst of change, you spark the flame,
Transforming compounds, never the same.
In the alchemist's hands, you work your charm,
Praseodymium, catalyst of transformative art.

Oh Praseodymium, element of allure,
In the world of science, you endure.
With your atomic charm, so unique and rare,
Praseodymium, a wonder beyond compare.

In lasers, you dazzle, emitting your light,
A dance of photons, a celestial sight.
Praseodymium, illuminating the night,
Guiding us forward, with your radiant might.

Oh Praseodymium, element divine,
In the realm of elements, you truly shine.
With your mysteries and wonders, forever you'll be,
Praseodymium, an enigmatic key.

TWENTY-THREE

WITHIN THEE

Praseodymium, element of boundless might,
In the realm of science, your secrets ignite.
With atomic grace, you capture our gaze,
Unfolding mysteries in ever-curious ways.

Oh Praseodymium, conductor of change,
In the alchemist's hands, you rearrange.
Catalyzing reactions, you spark the flame,
Praseodymium, a catalyst without blame.

In the realm of lasers, you radiate,
Casting a vibrant glow, a luminescent state.
Praseodymium, illuminating the night,
Guiding us forward with your radiant light.

Oh Praseodymium, element so rare,
With your magnetic charm, we stare.
Your presence, magnetic and strong,
Praseodymium, you can do no wrong.

In the depths of the periodic table, you reside,
A symbol of wonder, impossible to hide.
Praseodymium, an enigmatic key,
Unlocking the mysteries that lie within thee.

TWENTY-FOUR

SECRETS OF CHEMISTRY

In the realm of elements, a jewel so rare,
Praseodymium, you dazzle with flair.
With atomic grace, you command the stage,
A chemical marvel, beyond our gauge.

Oh Praseodymium, iridescent and bright,
In the depths of science, you ignite.
With magnetic allure, you draw us near,
A conductor of forces, crystal clear.

In lasers, you dance with ethereal grace,
Painting the canvas of light in space.
Praseodymium, a luminescent guide,
Leading us forward, side by side.

Oh Praseodymium, element of intrigue,
In the alchemist's hands, you work your mystique.

Catalyst of change, transforming the game,
Praseodymium, forever engraved in fame.
  With your unique properties, you amaze,
A symbol of curiosity that never decays.
Praseodymium, an enigmatic key,
Unleashing the secrets of chemistry.

TWENTY-FIVE

ELEMENT WITH SOUL

In the realm of elements, you stand tall,
Praseodymium, captivating us all.
A rare gem of the periodic table,
Your presence, a scientific fable.

With magnetic allure, you enthrall,
Praseodymium, conductor of all.
Guiding forces with your magnetic might,
Invisible threads woven in the night.

In lasers, you shimmer with vibrant grace,
Casting a green hue, filling the space.
Praseodymium, illuminating the way,
A beacon of light, in the darkest day.

A catalyst of change, you hold the key,
Unleashing transformations, for all to see.

Praseodymium, alchemist's delight,
Unveiling the secrets, shining so bright.
  Oh Praseodymium, element of wonder,
In the realm of science, forever you ponder.
With your atomic dance, a mesmerizing sight,
Praseodymium, shining with celestial light.
  In the vast universe, you claim your role,
Praseodymium, an element with soul.
A symbol of curiosity, in science you thrive,
Praseodymium, forever you'll strive.

TWENTY-SIX

ELEMENT OF ALLURE

Praseodymium, rare and divine,
In the realm of elements, you truly shine.
With atomic grace, you captivate,
A symphony of colors, an ethereal state.
    In lasers, you dance with vibrant flair,
Casting a spell, enchanting the air.
Praseodymium, conductor of light,
Guiding us through darkness with all your might.
    Oh Praseodymium, a catalyst profound,
In the alchemist's hands, miracles abound.
Transforming reactions with a magical touch,
Praseodymium, artistry that means so much.
    In the tapestry of science, you weave,
A puzzle piece that makes us believe.

Praseodymium, element of intrigue,
Unveiling mysteries, like an ancient league.
 A symbol of curiosity, you ignite,
Inquisitive minds, forever taking flight.
Praseodymium, element of allure,
Unveiling the secrets, forever pure.

TWENTY-SEVEN

PUZZLE TO EXPLORE

In the realm of elements, you shine so bright,
Praseodymium, a celestial light.
With magnetic allure, you captivate,
Drawing us closer, we cannot hesitate.
Oh Praseodymium, catalyst of change,
In the alchemist's hands, you rearrange.
Transforming reactions, with a magical spark,
Praseodymium, leaving a lasting mark.
In the world of lasers, you dance and glow,
A vibrant display, a captivating show.
Praseodymium, illuminating the way,
Guiding us forward, day by day.
Oh Praseodymium, element divine,
Your presence, so rare, forever will shine.
A symbol of curiosity, forever we'll seek,
Praseodymium, enigmatic and unique.

In the depths of science, you hold the key,
Unveiling the secrets, for all to see.
Praseodymium, a puzzle to explore,
A marvel of nature, that we adore.

TWENTY-EIGHT

DIVINE

In the realm of elements, you claim your throne,
Praseodymium, a jewel of your own.
With magnetic allure, you draw us near,
A conductor of forces, crystal clear.
    In the depths of lasers, you shimmer and gleam,
Casting a vibrant glow, a mesmerizing dream.
Praseodymium, a luminescent guide,
Leading us forward, side by side.
    Oh Praseodymium, element of grace,
In the alchemist's hands, you find your place.
Catalyst of change, transforming the game,
Praseodymium, forever etched in fame.
    With your unique properties, you amaze,
A symbol of curiosity that never decays.

Praseodymium, an enigmatic key,
Unlocking the mysteries that lie within thee.
   In the tapestry of science, you weave,
A puzzle piece that makes us believe.
Praseodymium, element of wonder,
Unveiling secrets, as lightning and thunder.
   With your presence, the world becomes aware,
Praseodymium, a gem beyond compare.
A symbol of progress, forever you'll shine,
Praseodymium, an element so divine.

TWENTY-NINE

PRASEODYMIUM, AN ELEMENT REVERED

In the realm of elements, you hold your sway,
Praseodymium, in shades of green and gray.
A conductor of light, with a magnetic charm,
You dazzle the world, like a celestial alarm.

Oh Praseodymium, catalyst of change,
In the alchemist's hands, you rearrange.
Transforming reactions with a mystical might,
Praseodymium, shining ever so bright.

In lasers, you dance with ethereal grace,
Painting the canvas of light in space.
Praseodymium, a luminescent guide,
Leading us forward, side by side.

With your atomic dance, a cosmic ballet,
Praseodymium, you mesmerize, we say.

A symbol of curiosity, a beacon of awe,
Unveiling the secrets of nature's grand law.
    Oh Praseodymium, element divine,
In the realm of science, you forever shine.
With your presence, the world's path is cleared,
Praseodymium, an element revered.

THIRTY

SCIENTIFIC DESIRE

In the realm of elements, you hold your sway,
Praseodymium, a jewel of the periodic array.
With atomic number fifty-nine, you stand tall,
A testament to nature's intricate call.

Praseodymium, a symphony of green,
Your hues emerge, a sight to be seen.
In the world of magnets, you play your part,
Aligning fields with your magnetic art.

A metal of rare earth, you hold the key,
To innovation and progress, for all to see.
Praseodymium, conductor of light,
Guiding us forward, shining so bright.

In the laboratory, you captivate,
Scientists amazed by your chemical state.

Praseodymium, an enigma, yet to be unveiled,
Unraveling mysteries, as science prevailed.

A symbol of curiosity, you beckon us near,
Praseodymium, your presence we revere.
In the grand tapestry of the universe's plan,
You shine with brilliance, like a celestial fan.

Oh Praseodymium, element of wonder and awe,
Forever you'll enchant, forever you'll draw.
With your unique properties, you inspire,
Praseodymium, a symbol of scientific desire.

THIRTY-ONE

INTRIGUING

Praseodymium, jewel of the periodic table,
A symphony of electrons, a chemical fable.
With atomic number fifty-nine, you reside,
In the realm of rare earth metals, your pride.

In the alchemist's hands, you come alive,
A catalyst of transformation, a vibrant dive.
Praseodymium, conductor of colors unseen,
Painting the world with a mesmerizing sheen.

Through the crystals, your essence unfurls,
Revealing a spectrum of iridescent swirls.
Praseodymium, a prism of endless hues,
Enchanting hearts, igniting muse.

In the depths of science, your secrets lie,
Unveiling the mysteries that make us sigh.

Praseodymium, a symbol of wisdom untold,
Guiding explorers as they venture bold.

Oh Praseodymium, element divine,
Your presence, a symbol of strength and shine.
In the grand tapestry of the universe's scheme,
You illuminate the path, like a celestial beam.

With every electron's dance, you inspire,
Praseodymium, a symphony that never tires.
A testament to the wonders science can find,
Praseodymium, forever intriguing, forever kind.

THIRTY-TWO

EVERY DISCOVERY

Praseodymium, an element of grace,
In the realm of chemistry, you find your place.
With atomic beauty, you captivate,
A symphony of electrons, orchestrating fate.

Your vibrant spectrum, a kaleidoscope delight,
Praseodymium, painting the world with light.
In the lab, scientists marvel at your power,
Unveiling secrets, hour by hour.

Oh Praseodymium, a gem of rare worth,
In the periodic table, a treasure unearthed.
With magnetic allure, you pull us close,
Guiding us on a journey, where knowledge flows.

A symbol of progress, you stand tall,
Praseodymium, inspiring one and all.

Your presence, a spark that lights the fire,
Igniting curiosity, igniting desire.

In the realm of elements, you shine bright,
Praseodymium, a beacon of scientific light.
With every discovery, a new chapter unfolds,
Praseodymium, a story yet to be fully told.

THIRTY-THREE

ZEST

    Praseodymium, a whisper in the earth's core,
A mystery waiting to be explored.
With atomic grace, you dance in the night,
A celestial symphony, shining so bright.
    In the realm of elements, you hold your reign,
Praseodymium, a catalyst for change.
Your magnetic charm, a force to behold,
Guiding the currents, untangling the fold.
    Oh Praseodymium, element of grace,
Within your essence, a captivating embrace.
From ancient lands to the present day,
You've left your mark in a remarkable way.
    In the laboratory's embrace, you reveal,
The secrets of matter, with an ethereal zeal.
Praseodymium, a symbol of innovation,
Fueling the fires of scientific creation.

With each electron's dance, a story unfolds,
Praseodymium, a tale waiting to be told.
In the realm of chemistry's curious quest,
You shine as a symbol of eternal zest.

THIRTY-FOUR

FOREVER INSPIRING

Praseodymium, a jewel of the periodic table,
Your radiance shines, a luminescent fable.
In the realm of elements, you hold a special place,
With your atomic structure, a cosmic embrace.
Oh Praseodymium, your magnetic allure,
Captivating scientists, their curiosity pure.
Unraveling the mysteries of your atomic dance,
Revealing the secrets with scientific advance.
A symbol of resilience, with a vibrant hue,
Praseodymium, you inspire, you renew.
In the world of alloys, you lend your strength,
Enhancing materials, pushing the length.
With your presence, innovation takes flight,
Praseodymium, a beacon of scientific light.
From lasers to magnets, you pave the way,
Guiding us forward, day by day.

Oh Praseodymium, element divine,
In the grand tapestry of science, you shine.
A testament to the wonders we explore,
Praseodymium, forever inspiring, forever more.

THIRTY-FIVE

SCIENTIFIC WAYS

Praseodymium, a jewel of the periodic table,
In your presence, science and wonder enable.
With atomic grace, you captivate,
Unveiling secrets that scientists elucidate.
Praseodymium, symbol of strength and might,
Guiding us through the realms of knowledge's light.
A catalyst for change, you inspire,
Fueling innovation with a burning desire.
In the laboratory's hallowed ground,
Your mysteries unravel, profound.
Oh Praseodymium, element of rare earth,
In your composition, discoveries find rebirth.
From lasers to magnets, you lend your hand,
Pushing boundaries, defying the bland.
With each electron's dance, a symphony plays,
Praseodymium, the conductor of scientific ways.

In the grand tapestry of elements untold,
You shine brightly, a story yet to unfold.

ABOUT THE AUTHOR

Walter the Educator is one of the pseudonyms for Walter Anderson. Formally educated in Chemistry, Business, and Education, he is an educator, an author, a diverse entrepreneur, and he is the son of a disabled war veteran. "Walter the Educator" shares his time between educating and creating. He holds interests and owns several creative projects that entertain, enlighten, enhance, and educate, hoping to inspire and motivate you.

Follow, find new works, and stay up to date
with Walter the Educator™
at WaltertheEducator.com

www.ingramcontent.com/pod-product-compliance
Lightning Source LLC
LaVergne TN
LVHW020133080526
838201LV00117B/3739